A New True Book

CONTINENTS

By Dennis B. Fradin

CHILDRENS PRESS ®

CHICAGO

Seacow Head, Prince Edward Island, Canada

Dedication: For Leo Cohen

Library of Congress Cataloging-in-Publication Data

Fradin, Dennis B.
   Continents.

   (A New true book)
   Includes index.
   Summary: Defines and enumerates the continents of
the earth, describing their countries, geography,
geology, origin, movement, and possible future.
   1. Continents—Juvenile literature. [1. Continents]
I. Title.
G133.F73 1986   910'.09141   86-9580
ISBN 0-516-01291-6

PHOTO CREDITS
© Cameramann International, Ltd.—43
(left)

© Virginia Grimes—37 (bottom left)

Journalism Services: © Ellen H.
Przekop—10 (bottom left)

Courtesy NASA—4

John Forsberg—5

Nawrocki Stock Photo: © Janet Davis—
10 (bottom right)

Odyssey Productions, Chicago: © Robert
Frerck—20 (right), 31 (right)

Photri—40; © J. Allen Cash—10 (top
right)

The Photo Source, International—Cover

H. Armstrong Roberts:
© Camerique—10 (top left), 39
© E.R. Degginger—31 (left)
© R. Lamb—37 (bottom right)

Root Resources: © Bill Glass—6 (top)

Tom Stack & Associates: © Brian
Parker—18

© Tass/Sovfoto—19 (left)

Valan Photos:
© Kennon Cooke—6 (bottom)
© Jean-Marie Jro—43 (right)
© S.J. Krasemann—19 (right)
© Joseph R. Pearce—2

Albert R. Magnus—9, 12, 14, 15, 16, 17,
18, 20, 21, 23, 25, 26, 29, 33, 35, 37
(top), 42

Cover: A panorama of the
        Swiss Alps

TABLE OF CONTENTS

Earth from space

WHAT ARE CONTINENTS?

We live on the lovely planet Earth. From space, most of Earth looks blue. That is because more than two thirds of its surface is covered by water.

Western Hemisphere

Eastern Hemisphere

Snow cap and iceberg (above) along the coast of Greenland
Thousand Islands in the St. Lawrence River (below)

Less than one third of Earth's surface is land. The smallest pieces of land are called islands. The tiniest islands are just a few feet across. The largest island is Greenland, which spreads over more than 800,000 square miles.

Earth's largest landmasses are called continents. Continents cover millions of square miles.

HOW MANY CONTINENTS ARE THERE?

Earth is usually said to have seven continents. From biggest to smallest, they are:

Continent	Approximate square miles
1. Asia	17,000,000
2. Africa	12,000,000
3. North America	9,000,000
4. South America	7,000,000
5. Antarctica	5,000,000
6. Europe	4,000,000
7. Australia	3,000,000

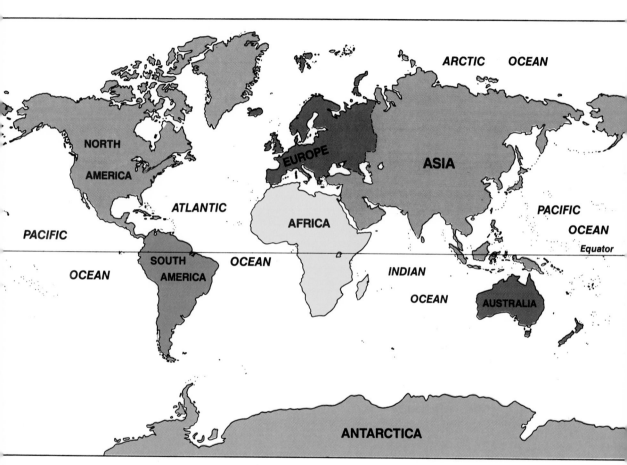

Australia is an island continent.

Children from China (top left), Bali, Indonesia (top right), Ireland (below), and Spain (right)

For centuries, most people have considered Europe a separate continent from Asia. Perhaps this is because most Asians differ in appearance from most Europeans. But a map shows that Europe is attached to Asia. No body of water separates them. Because of this, many geographers (experts on Earth's surface features) consider Europe and Asia a single continent. They call it Eurasia. Viewed in this way, the world has only six continents:

Continent	Approximate square miles
1. Eurasia	21,000,000
2. Africa	12,000,000
3. North America	9,000,000
4. South America	7,000,000
5. Antarctica	5,000,000
6. Australia	3,000,000

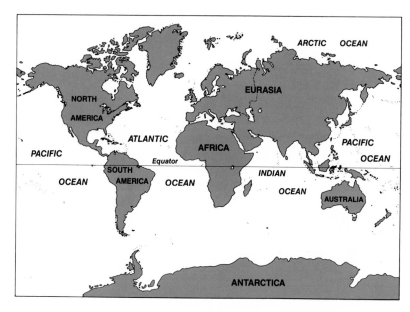

COUNTRIES
AND CONTINENTS

Some people confuse continents with countries. They are two very different things. A country is a political unit. Countries often have man-made borders. Continents, on the other hand, were made by natural processes.

Antarctica has no independent countries. That is because it is so cold

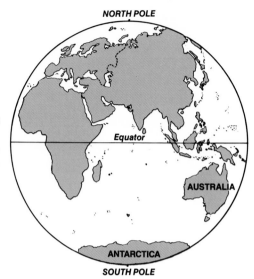

An imaginary line called the equator separates the Earth into the Northern Hemisphere and the Southern Hemisphere

there! Australia has just one country on it, also called Australia. Each of the other continents has many countries. For example, North America includes the countries of Canada, the United States, Mexico and Central America.

GREENLAND

ALASKA
(U.S.)

CANADA

UNITED   STATES

PACIFIC

OCEAN

ATLANTIC

OCEAN

Gulf of Mexico

MEXICO

Central America

15

PHYSICAL FEATURES OF CONTINENTS

There are several types of physical features on the continents. Among the main ones are mountains, plains, lakes, and rivers.

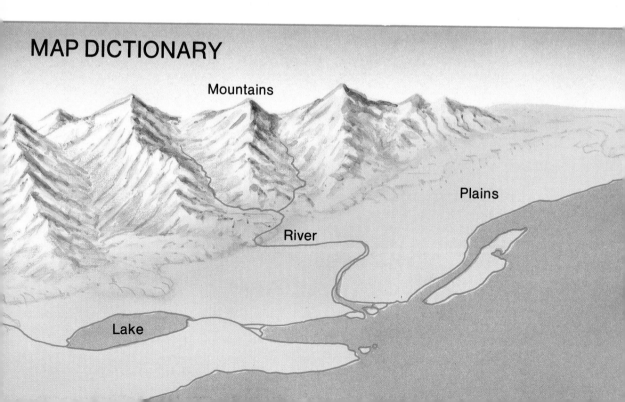

MAP DICTIONARY

Mountains

Plains

River

Lake

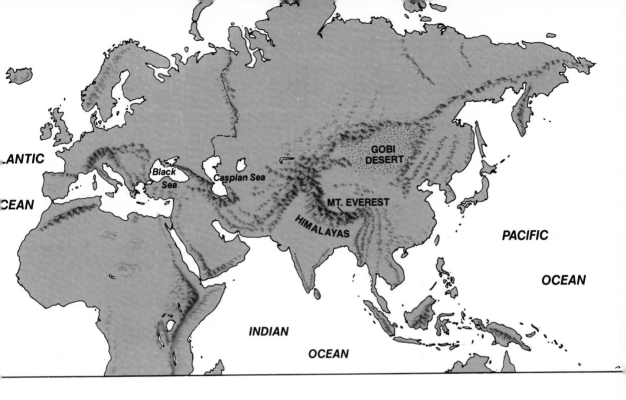

Every continent has
major mountain ranges.
Earth's tallest mountain is
Mount Everest, in Asia. It
rises about 29,000 feet—
more than five miles—
above sea level. Every
continent also has vast

Wheat harvest in the Great Plains of North America

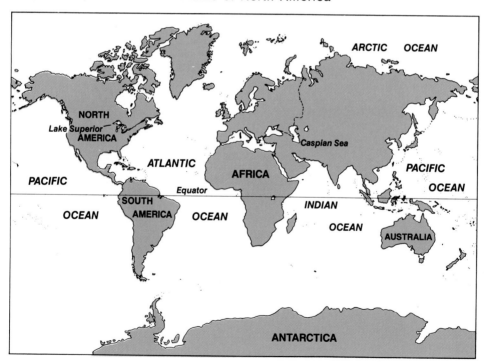

A 124-mile elevated road (left) crosses this section of the Caspian Sea. Lake Superior (above) is one of the Great Lakes in North America.

stretches of nearly level land called plains.

Lakes can be found on all the continents, too. The Caspian Sea, a saltwater lake lying in both Europe and Asia, is Earth's largest lake. The largest freshwater lake is North America's Lake Superior.

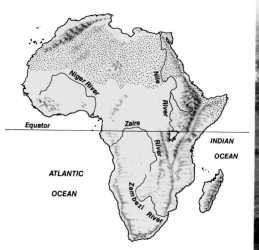

Bridges cross the Nile River in Cairo, Egypt.

All of the continents but Antarctica have rivers flowing across the land. The longest river, the Nile in Africa, is 4,145 miles long. Antarctica would have rivers, too, but most of its water is locked up as ice and snow.

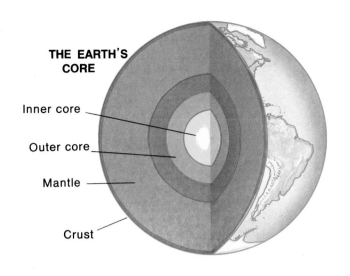

THE EARTH'S CORE

Inner core

Outer core

Mantle

Crust

THE STRUCTURE OF CONTINENTS

Our planet Earth has an outer skin, called the crust, that is made of rock.

Beneath the oceans, Earth's crust is only about five miles thick. The crust has only one layer of rocks there. Where the continents are, there are two layers of rocks. This extra layer helps make the continents higher than the oceans. The crust stretches down about twenty miles beneath the continents.

The continents do not end where the ocean meets the land. Instead, the continents extend offshore for at least a few miles. The underwater part of a continent is known as

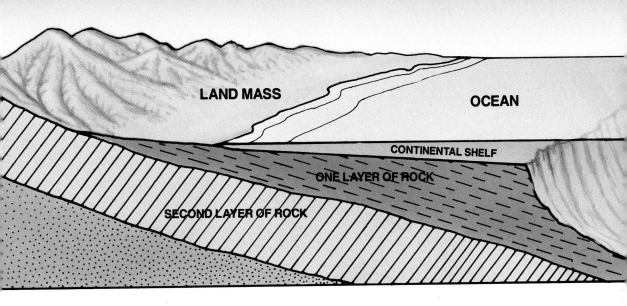

LAND MASS

OCEAN

CONTINENTAL SHELF

ONE LAYER OF ROCK

SECOND LAYER OF ROCK

the continental shelf.
Depending on their
location, continental
shelves vary in width from
a few miles to a few
hundred miles.

The continental shelves
slope gently downward.
The water atop the
continental shelves never

gets very deep, however.
At its deepest, the ocean
above the continental
shelves is only about six-
hundred-feet deep.

At the point where a
continental shelf ends,
there is a sudden steep
downward slope, called the
continental slope. By the
time the continental slope
finishes its drop, the ocean
is several miles deep. At
that point the vast ocean
bed begins.

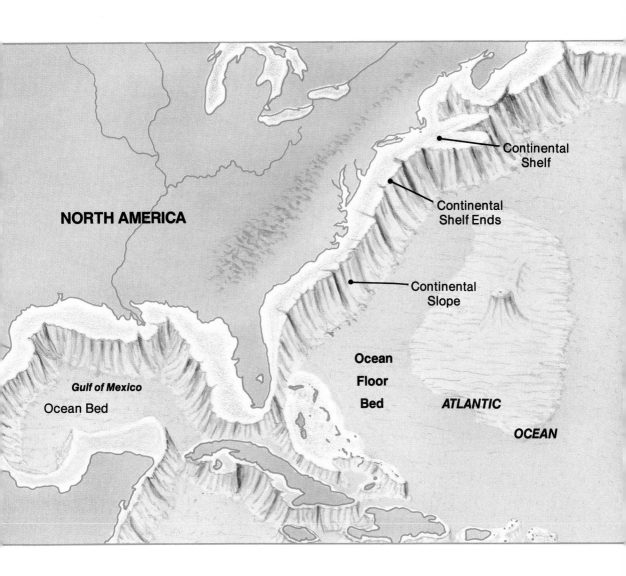

NORTH AMERICA

Continental
Shelf

Continental
Shelf Ends

Continental
Slope

Ocean

Floor

Bed

Gulf of Mexico
Ocean Bed

ATLANTIC

OCEAN

25

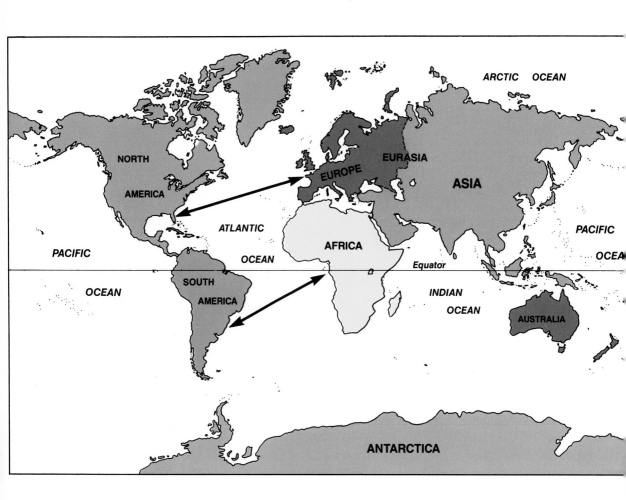

ARCTIC   OCEAN

NORTH

AMERICA

EUROPE

EURASIA

ASIA

ATLANTIC

PACIFIC

OCEA

PACIFIC

OCEAN

AFRICA

Equator

OCEAN

SOUTH

AMERICA

INDIAN

OCEAN

AUSTRALIA

ANTARCTICA

26

HOW WERE CONTINENTS CREATED?

Compare the continents' coasts on a map or globe. In many places, the continents could be fitted together like pieces in a jigsaw puzzle. For example, the eastern coast of South America would fit into Africa's western coast. Northeastern North America would fit neatly into Europe's northwestern coast.

Geologists (scientists who study Earth's insides) have a theory to explain this. They say that hundreds of millions of years ago, there was just one continent, which they call Pangaea. Pangaea took up about one third of Earth's surface. About 200 million years ago, Pangaea is thought to have split in two. The two sections spread apart at a rate of perhaps an inch per year.

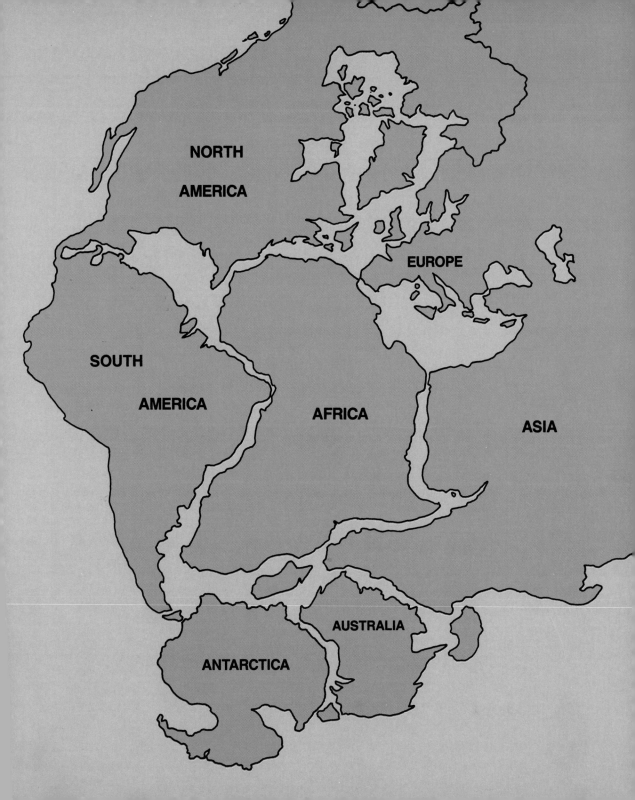

Over millions of years
these two big continents
split into the continents we
know today. Over many
more millions of years they
moved into their current
positions.

There is evidence to
support this theory. Fossils
(remains of plants and
animals) provide some
evidence. At places where
continents may once have
been joined, certain plants
and animals developed at
the same time. This might

Kangaroo with its baby, called a joey (left), and
the fuzzy koala bear (right) are found only in Australia,
an island continent.

have occurred on far-apart
continents. But it is more
likely they evolved on a
single landmass which
later broke apart.

CONTINENTAL DRIFT

The slow movement that is said to have caused Pangaea to slowly break and spread apart over millions of years is called continental drift.

Geologists have a theory to explain continental drift. They say that Earth's outer shell is made of a number of separate pieces called plates. There are seven

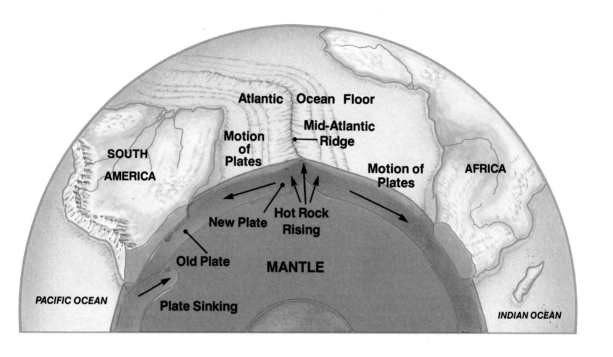

Atlantic Ocean Floor

Mid-Atlantic
Ridge

Motion
of
Plates

SOUTH
AMERICA

Motion of
Plates

AFRICA

New Plate

Hot Rock
Rising

Old Plate

MANTLE

PACIFIC OCEAN

Plate Sinking

INDIAN OCEAN

big plates and many
smaller ones. About sixty
miles thick, the plates
stretch down through the
crust into the hot rock
below.

These plates are believed to move a little each year. As they do, several things occur. For one, they move the continents and the ocean bottoms a short way each year. The movement of the plates probably broke Pangaea apart and moved the continents into their current positions.

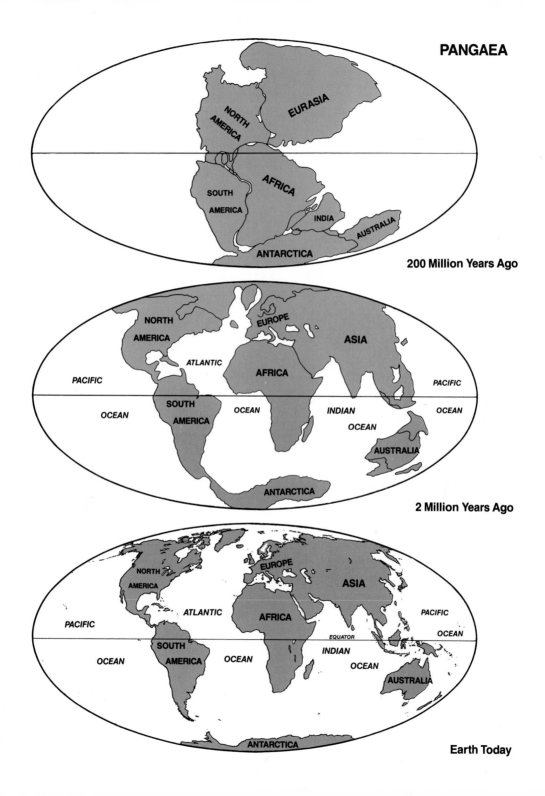

PANGAEA

NORTH AMERICA

EURASIA

SOUTH AMERICA

AFRICA

INDIA

AUSTRALIA

ANTARCTICA

200 Million Years Ago

NORTH AMERICA

EUROPE

ASIA

PACIFIC OCEAN

ATLANTIC

AFRICA

SOUTH AMERICA

OCEAN

INDIAN OCEAN

PACIFIC OCEAN

AUSTRALIA

ANTARCTICA

2 Million Years Ago

NORTH AMERICA

EUROPE

ASIA

PACIFIC

ATLANTIC

AFRICA

SOUTH AMERICA

OCEAN

EQUATOR

INDIAN OCEAN

PACIFIC OCEAN

OCEAN

AUSTRALIA

ANTARCTICA

Earth Today

As the plates move about, underground rocks sometimes snap. This causes earthquakes. Sometimes plates collide and one plate is pushed beneath another into Earth's very hot insides. When this occurs, some rock melts. The melted rock rises to Earth's surface, forming volcanoes.

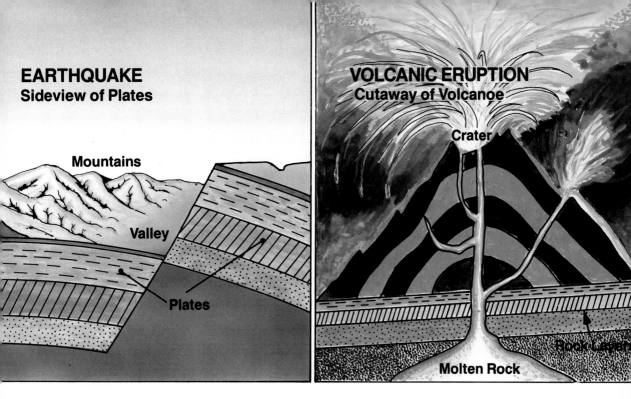

EARTHQUAKE
Sideview of Plates

Mountains

Valley

Plates

VOLCANIC ERUPTION
Cutaway of Volcanoe

Crater

Rock Layers

Molten Rock

Santa Maria volcanic cone (below left) at Lake Atitlan, Guatemala. The smoking crater of the still-active Mt. Saint Helens (below right) in Washington State.

ARE THERE
LOST CONTINENTS?

For ages, people have talked about "lost continents," great land areas that are said to have disappeared. The best-known story is about Atlantis, which supposedly sank into the sea thousands of years ago. Even the famous Greek philosopher Plato described Atlantis.

Village of Dei on Santorini

Some scientists think
that Atlantis really was the
island of Santorini (now
called Thira) in the
Aegean Sea. In about
1500 B.C. part of Santorini

was destroyed by a volcano and disappeared into the sea.

No piece of land the size of a continent has ever been found beneath the sea. But to this day, stories about Atlantis and other "lost continents" are still told.

THE FUTURE
OF THE CONTINENTS

The continents are thought to be still drifting. We can't feel the drift because it is only about an inch per year. Over millions of years, however, those inches add up to thousands of miles.

Scientists say that millions of years from now Africa and South America will be much farther apart.

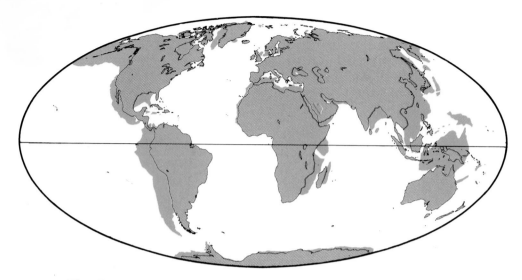

The Earth's appearance 25,000,000 years from now. By then Australia will have moved closer to Asia.

Europe and North America will also move apart, but Australia and Asia will move close together. California may even tear away from the North American continent and

Oceans carved the coastlines of continents and giant mountains, caused by the massive earthquakes millions of years ago, are landmarks on modern maps.

become an island! Maps won't have to be altered for a long time, however, because these changes will occur over millions of years.

CONTINENTS and COUNTRIES

AFRICA

Name	Capital	Name	Capital	Name	Capital
Algeria	Algiers	Guinea	Conakry	Reunion	Saint-Denis
Angola	Luanda	Guinea-Bissau	Bissau	(French)	
Benin	Porto-Novo	Ivory Coast	Abidjan	Rwanda	Kigali
Botswana	Gaborone	Kenya	Nairobi	São Tomé and	São Tomé
Burkina Faso	Ouagadougou	Lesotho	Maseru	Principe	
Burundi	Bujumbura	Liberia	Monrovia	Senegal	Dakar
Cameroon	Yaoundé	Libya	Tripoli	Seychelles	Victoria
Canary Islands	Santa Cruz	Madagascar	Antananarivo	Sierra Leone	Freetown
(Spanish)	de Tenerife;	Malawi	Lilongwe	Somalia	Mogadishu
	Las Palmas	Mali	Bamako	South Africa	Cape Town;
Cape Verde	Praia	Mauritania	Nouakchott		Pretoria;
Central African	Bangui	Mauritius	Port Louis de		Bloemfontein
Republic			Gran Canaria	Sudan	Khartoum
Chad	N'Djamena	Madeira Islands	Funchal	Swaziland	Mbabane
Comoros	Moroni	(Portugese)		Tanzania	Dar es Salaam
Congo	Brazzaville	Morocco	Rabat	Togo	Lomé
Djibouti	Djibouti	Mozambique	Maputo	Tunisia	Tunis
Egypt	Cairo	Namibia (South	Windhoek	Uganda	Kampala
Equatorial	Malabo	West Africa)		Western Sahara	None
Guinea		(controlled by		(disputed)	
Ethiopia	Addis Ababa	South Africa)		Zaire	Kinshasa
Gabon	Libreville	Niger	Niamey	Zambia	Lusaka
Gambia	Banjul	Nigeria	Lagos	Zimbabwe	Harare
Ghana	Accra				

ANTARCTICA—an island country

ASIA

Name	Capital	Name	Capital	Name	Capital
Afghanistan	Kabul	Japan	Tokyo	Qatar	Doha
Bahrain	Manama	Jordan	Amman	Saudi Arabia	Riyadh
Bangladesh	Dhaka	Kampuchea	Phnom Penh	Singapore	Singapore
Bhutan	Thimphu	Korea, North	Pyongyang	Soviet Union	Moscow
Brunei	Bandar Seri	Korea, South	Seoul	Sri Lanka	Colombo
	Begawan	Kuwait	Kuwait	Syria	Damascus
Burma	Rangoon	Laos	Vientiane	Taiwan	Taipei
China	Peking	Lebanon	Beirut	Thailand	Bangkok
Cyprus	Nicosia	Macao	Macao	Turkey (Asian)	Ankara
Hong Kong	Victoria	Malaysia	Kuala Lumpur	United Arab	Abu Dhabi
(British)		Maldives	Male	Emirates	
India	New Delhi	Mongolia	Ulan Bator	Vietnam	Hanoi
Indonesia	Jakarta	Nepal	Katmandu	Yemen (Aden)	Aden
Iran	Teheran	Oman	Muscat	Yemen (Sana)	Sana
Iraq	Baghdad	Pakistan	Islamabad		
Israel	Jerusalem	Philippines	Manila		

AUSTRALIA—an island country

EUROPE

Name	Capital	Name	Capital	Name	Capital
Albania	Tiranë	Greece	Athens	Portugal	Lisbon
Andorra	Andorra	Hungary	Budapest	Romania	Bucharest
Austria	Vienna	Iceland	Reykjavik	Russia	Moscow
Belgium	Brussels	Ireland	Dublin	(European)	
Bulgaria	Sofia	Italy	Rome	San Marino	San Marino
Czechoslovakia	Prague	Liechtenstein	Vaduz	Spain	Madrid
Denmark	Copenhagen	Luxembourg	Luxembourg	Sweden	Stockholm
Finland	Helsinki	Malta	Valletta	Switzerland	Bern
France	Paris	Monaco	Monaco	Turkey	Ankara
Germany (East)	Berlin (East)	Netherlands	Amsterdam	(European)	
Germany (West)	Bonn	Norway	Oslo	Vatican City	
Great Britain	London	Poland	Warsaw	Yugoslavia	Belgrade

NORTH AMERICA

Name	Capital	Name	Capital	Name	Capital
Anguilla	"The Valley"	Grenada	Saint George's	St. Lucia	Castries
Antigua and Barbuda	St. John's	Guadeloupe	Basse-Terre Island	St. Pierre and Miquelon	St. Pierre
Bahamas	Nassau	Guatemala	Guatemala City	St. Vincent and the Grenadines	Kingstown
Barbados	Bridgetown	Haiti	Port-au-Prince		
Belize	Belmopan	Honduras	Tegucigalpa	Trinidad and Tobago	Port-of-Spain
Bermuda	Hamilton	Jamaica	Kingston		
Canada	Ottawa	Martinique	Fort-de-France	Turks and Caicos Islands	Grand Turk
Cayman Islands	Georgetown	Montserrat	Plymouth		
Costa Rica	San Jose	Mexico	Mexico City	United States	Washington, D.C.
Cuba	Havana	Netherlands Antilles	Wilemstad	Virgin Islands (U.S.)	Charlotte Amalie,
Dominica	Roseau				
Dominican Republic	Santo Domingo	Nicaragua	Managua	Virgin Islands (British)	Road Town
		Panama	Panama City		
El Salvador	San Salvador	Puerto Rico	San Juan		
Greenland		St. Christopher and Nevis	Basseterre		

SOUTH AMERICA

Name	Capital	Name	Capital
Argentina	Buenos Aires	French Guiana (Overseas department of France)	
Bolivia	La Paz; Sucre	Guyana	Georgetown
Brazil	Brasília	Paraguay	Asunción
Chile	Santiago	Peru	Lima
Colombia	Bogotá	Suriname	Paramibo
Ecuador	Quito	Uruguay	Montevideo
Falkland Islands (British dependency)	Stanley	Venezuela	Caracas

WORDS YOU SHOULD KNOW

B.C.—before Christ

coast(KOHST)—the land along a large body of water

continent(KAHN • tih • nent)—one of Earth's largest landmasses

continental drift(KAHN • tih • nen • til DRIFT)—the theory that continents slowly move

continental shelf(KAHN • tih • nen • til SHELF)—an extension of a continent which slopes down into the ocean

continental slope(KAHN • tih • nen • til SLOAP)—the underwater, very steep, outermost part of a continent which slopes down to the ocean bed

country(KUN • tree)—a nation with its own government and, often, man-made borders

crust(KRUST)—the outer skin, between several and twenty miles thick, of Earth

Earth(ERTH)—the planet on which we live

earthquake(ERTH • kwaik)—a shaking of the ground thought to be caused by the movement of Earth's plates

Eurasia(yoo • RAY • zhuh)—the name given to Europe and Asia when they are considered to be one continent

fossils(FAW • silz)—remains of plants and animals that lived long ago

geographer(gee • OG • gra • fer)—an expert on the features of Earth's surface

geologist(gee • OL • oh • gist)—a scientist who studies Earth's insides

island(EYE • land)—a small piece of land surrounded by water

million(MILL • yun)—a thousand thousand (1,000,000)

Pangaea(pan • GEE • ah)—the vast continent thought to have once comprised all of Earth's land

plain(PLAYNE)—relatively level land

planet(PLAN • it)—a large object that orbits a star

plates (of Earth) (PLAITZ) — the massive pieces comprising Earth's outer shell, which are thought to cause continental drift

theory (THEER • ree) — an unproven idea that attempts to explain something

thousand (THOW • zend) — ten hundred (1,000)

volcano (vol • KAY • noh) — an opening in Earth's crust through which liquid rock sometimes erupts; the mountain that builds up around the opening is also called a volcano

INDEX

About the author

Dennis Fradin attended Northwestern University on a partial creative scholarship and graduated in 1967. His previous books include the Young People's Stories of Our States series for Childrens Press, and Bad Luck Tony for Prentice-Hall. In the True book series Dennis has written about astronomy, farming, comets, archaeology, movies, space colonies, the space lab, explorers, and pioneers. He is married and the father of three children.